KEEPING ALPACAS
Secrets from the field

Helen Rose

ISBN 978-1482045994

DISCLAIMER

The information in this book is correct to the best of our knowledge. The author and the publisher do not accept any legal responsibility or liability for any errors or omissions that may have been made in this book.

The author and the publisher do not accept any legal responsibility or liability for any injury, damage or loss, or adverse outcome of any kind incurred as a result of the use of the material contained in this book or reliance upon it. If in doubt about any aspect of keeping alpacas readers are advised to seek professional advice.

Consult your veterinarian about the general health of your alpacas, to diagnose symptoms that may need medical attention and for advice on routine treatments that alpacas may require.

Acknowledgements

Thank you to all my friends and family. I could not have done it without you. Especially thanks to Mum, Duncan, Lisa, Lindsay, Kim, Jeanette and Sara.

Amazing Alpacas

Alpacas are amazing! They are hardy, look gorgeous and even smell nice, a bit like a wheat biscuit. They are gentle, placid and relaxing to be with as they usually move slowly to conserve energy. They have a cheerful appearance with large dark eyes and an upturned mouth. Alpacas are surprisingly curious. If you stand gazing at the alpacas, they will be just as interested in you.

Alpacas produce a luxury fibre and are bred to sell as livestock or pets. They are successfully farmed both on a large commercial scale and on smallholdings.

Alpaca wool is as soft as cashmere, similar to silk in strength and very light in weight. The fleece comes in seven colours and twenty-two shades. It is lovely wool to handle, spin and knit.

Alpacas make excellent pets. They are easy to train and usually very gentle with children. Alpacas are also gentle on your land because they have a soft padded foot. As well as being great lawn mowers alpacas are an effective protector of livestock on the farm. They love babies and will protect lambs when kept with sheep, by chasing away predators such as foxes.

They are a quiet animal really but they do communicate with sound. Mothers cluck quietly to their young babies. Alpacas sound the alarm by making a loud intermittent call. When unsure they make a humming sound and each alpaca hums a different note.

History of the Alpaca

Alpacas, llamas, guanaco and vicuna are all members of the South American camelid (or camel) family. Eleven million years ago their earliest ancestors appeared in North America, and later moved to South America and Asia. These Asian animals became the old world camels, the bactrian and dromedary that we know today.

About seven thousand years ago the alpaca was domesticated from the vicuna. The alpaca fibre was prized by the Incas and worn by the nobility. The Inca empire was remarkably successful in refining the alpaca through controlled breeding, according to archaeologists; the Inca alpacas had a fibre quality superior to modern day animals.

Llama at Machu Picchu photo by Jeanette Last

When the Spanish invaded South America in the 16th century they introduced merino sheep in place of the alpaca herds. Alpacas moved to the high elevations on the Andes altiplano where sheep and cattle cannot survive. Here the alpaca adapted to the sparse vegetation, the extremes of temperature and reproduced without a refined breeding program.

Alpacas Today

Today, in southern Peru there are large alpaca farms with controlled husbandry. Often based on the altiplano, the alpacas live on spare vegetation and the temperature ranges from minus ten to over twenty degrees Celsius. The alpacas are not fenced in but allowed to roam during the daylight hours. They are brought in at night to stone corals for safe keeping from predators and the extremes of wind and weather. Intelligent and helpful they come in easily with just one or two people herding them. Selective breeding does occur but often the parentage is not recorded. At one time the Peruvian mills paid on the weight of fleece rather than fineness and llamas were crossed with alpaca to give heavier fleeces. When buying an alpaca be careful of animals showing traits of llama genes. They may look like a very tall alpaca or have banana shaped ears, and maybe a high set tail.

There are two types of alpaca, the huacaya and the suri. The majority of alpacas in the UK are huacaya and look like a teddy bear. The suri has long ringlets of hair that hang downwards. Suri are more flighty than the huacaya in nature and need more care in cold weather because their coat hangs downwards exposing the spine. Male alpacas are called machos, females are called hembras, yearlings are called tuis and babies are called cria.

The English textile merchant Sir Titus Salt discovered the qualities of alpaca fibre. Queen Victoria wore dresses made of alpaca cloth and an alpaca gentleman's coat was very popular. Sir Titus Salt's mill and Saltaire, the model village built for his workers in West Yorkshire England, is a world heritage site. Today the alpaca is farmed in South America, also in North America, Canada, Australia, New Zealand, Europe and England.

Alpaca Experiences

Years ago I was mad about horses and kept two on our smallholding. All that mucking out had an effect and I got a bad back. I also grew up a bit and realised the risks of riding. To cut a long story short two good homes were found for the equines and I had an empty paddock.

Two years passed, I did a lot of research on business opportunities for a few acres of land. The answer was alpacas, a good business opportunity and they are the easiest livestock to keep. My learning curve began when I started looking after my own animals and a new business with alpacas.

Time with the alpacas

I kept a diary and noted down some of my experiences to help other new owners with their alpacas. My diary started when I wrote about birthing the babies. From then on I noted down things that were useful and then I kept a more complete record of things that happened in a whole year.

I hope this record will be useful to anyone wanting to keep alpacas on a smallholding. So here is my year with alpacas, starting with birthing the babies and including notes of interest.

Birthing Cria

A year on from the birth of our first two alpaca cria and I had forgotten a lot. So I have written myself a diary about the birthing of this year's cria and some general notes on what to expect. I will refer to this in the future!

Newborn alpaca

I monitored our expectant mums. I knew they had gone into labour because they were humming a lot, going to the toilet very frequently and had udders very full of milk! Both babies were born easily and while I was not looking. I was told that if paddocks are in balance mineral wise and alpacas are fed cider vinegar regularly that they will have no trouble unpacking!

Once the cria is born, rub dry with some clean towels. The cord can be washed with soapy water and dipped in iodine solution. Lamb cord spray can be used instead and this can be done two to three times during the first day.

This is very important because pathogens can get into the babies system and cause infections. If the cria is limping or has a swollen eye this could be caused by an infection, don't take chances, call the vet.

Mum will usually pass the placenta within the hour (up to 3 hours), check it's all complete before disposing of it. I dig a little hole and bury it.

I have sprayed the cord

The cria needs to be standing up in 30 minutes (up to 1.5 hours) and feeding in 45 minutes (to 3 hours). Check the cria is really sucking and not just licking around the milk bar. Also if the cria is slow to feed, we give reconstituted powdered colostrum and this gives the cria a boost. This is best given in the first 5 hours after the cria is born.

Make sure the cria has passed the meconium, this should happen in the first day. Our cria was straining to pass something on his second day and nothing seemed to come out.

Luckily for me that day we were hosting our local alpaca group and I had the first hand advice of my more experienced alpaca breeder friends. They said he seemed a little shaky, however, alpaca babies are often very tired on their second day. They also said that all the meconium needs to come out before the cria can feed properly.

The solution is to give the cria an enema. I got out a large syringe. Using warm water with a bar of soap just swished around in it, I part filled the syringe and did the deed. Cria ran away afterwards, went to the toilet and then jumped for joy followed by galloping round the paddock! It was a fantastic to watch him so obviously feeling better.

Our yearling female was trying to play with the babies almost as soon as they were born. She was very gently running the pad of her foot down their back. Also once they were up and walking she was raising a foreleg next to them as if to jump on their back. I separated her for a day and consulted with other alpaca people. They all said she should be fine and just wants to be 'Auntie' to them. I tried her with them and she was perfectly behaved and the babies even chased her around the paddock.

A few days later, one of the mum alpacas had a good roll in the dust patch, while I was picking up manure. Looking at where she had rolled a lot of opaque white jelly type discharge was lying on the ground. Worried about infection I asked the vet to call and check the mums over. The vet checked temperature had a good look at the bottom end of things and declared the mums fit and said that the discharge was a normal part of the drying-up process for mum alpacas.

As the babies were born in the first part of September I was advised to put a coat on the babies, at night, for the first week and also to weigh the baby everyday to check they were gaining. Newborn babies should weigh between 5.5 to 10kg. After 7 days the cria coat was more like a waistcoat on the largest cria so I decided to let her go without. The other cria is much smaller in size so he will have his on for a bit longer. The babies gained 2kg in the first week.

Having two more alpaca babies has been a lovely experience and now I have my birthing diary and notes I will be better prepared for next years babies.

Alpaca and her cria

January

It has been snowy and cold on and off since before Christmas. The thermometer says the temperature fell to minus eight Celsius overnight. Last night, I fussed a lot about where to leave the animals. The alpacas can become stressed when they are shut in and changing location just before dusk may also stress them. They all had access to shelter or areas that are snow free. So in the end I let the alpacas choose where to sleep.

Taking shelter from the snow

11

In the morning I went out at 7am. The females were asleep under a tree, no snow here. The two boys were in their shelter and in a sheltered spot behind the shed. The shed has an overhang and I have put a pallet at right angles to this. The shed and pallet stop wind, rain and snow from the North and West.

I have five females, two cria and two adult males. I am putting out food twice a day for the boys and females. The females are happily eating three slices of hay a day and there is plenty of extra hay under cover. The females have free access to the stable and the stable keeps their hay dry, as well as providing shelter and shade. Luckily I had stocked up with bags of alpaca mix and alfalfa before the big freeze began.

Correct Halter Fitting

Alpacas can only breathe through their nose and the halter must fit on the bony part of the nose. If the nose-band droops, onto the soft cartilage of the nose the alpaca may panic and be difficult to handle. Nylon head-collars can stretch with the body heat, so always fasten the head collar snugly around the back of the alpaca's neck, near the ears.

Correctly fitted halter

12th Jan

Today the alpacas had their vitamin A, D and E supplement. This comes in a gel and which is given every four weeks from September to March. Remember to order this in August ready for the winter. This vitamin gel is in a tube a bit like the ones, from the DIY store, for grouting around the bath. It has the same type of applicator gun as well. Today I herded all the females into the tiny catch pen and then gave them the gel. Normally I would move them one at a time into the pen. However today they were very happy being dosed as a group. I will definitely do it this way next time. I moved them using two long white sticks made from plastic tubes. I think they are sold as conduit for wires. Anyway, the sticks being white, the alpacas see them very easily and I use them to indicate where I want the group to move to. This also works extremely well, it's a bit like having two extra long arms!

To administer the vitamin gel I hold the alpaca's head in my left arm. Using my left hand to hold under the chin, I then guide the nozzle into the alpaca's mouth at the side (no teeth here) tilt the head slightly back and pull the trigger on the applicator which is in my right hand. We usually repeat this, every four weeks, from September through to March, when we have fewer hours of daylight hours in the UK.

Vitamins

I can feed each alpaca separately, putting their vitamins into their feed and they gobble up the feed with the vitamins attached.

For adults I put the feed in a skip and add the dose of vitamin paste (which contains A, D, E and B12). I then rub in the vitamin paste into the feed a bit like making pastry. One of my alpacas likes the taste and she comes over and tries to suck the paste out of the tube!

February

Today we are treating the alpacas for mites, by using an injection. We are treating all the alpacas because one has a case of ear mites that I cannot shift. I have tried using a tiny amount of mite medicine, topically in the ear. I have also tried oil used for treating mites. The oil seemed to give a reaction; she was shaking her head and rubbing her ears. On closer inspection the mites seemed to have made the ear edges crusty and hard. I have used a dog flea product to kill external mites. Once the mites were dead, some of the fur around the ears fell out and I used zinc and castor oil cream to help with the skin condition.

However, one other girl seemed to have mites, she was rubbing and scratching her ears and shaking her head a bit. Following injecting mite medicine both females then shook heads a lot but they settled after a few days.

Body Score

When a huacaya has grown a lot of fleece they look very round and fluffy. It is hard to know by looking at them whether they are thin or fat.

To check the condition, I feel the back of the alpaca, eight inches along the spine from the neck. Feeling the back is called body scoring. If the spine feels sharp they are underweight. If the back feels flat they are very overweight. Ideally an alpaca's back should feel padded over the spine with a flat slope going down either side. The slope should be neither: convex, too fat, or concave, too thin. Checking the animal's weight allows their feed to be adjusted according.

Normal Behavior

I watch my alpacas for some time each day. I know the normal behavior of each alpaca. Alpacas are stoic animals and mask signs of illness. If an alpaca is sick I can tell because they are not behaving normally.

Alpacas lie down and chew the cud and they often do this in the morning. Alpacas like to groom themselves by rolling in a dust bath. They will paw the ground before lying down and thrashing around rolling over to do both sides. All members of the herd will like to use the same area to roll. They often roll one after the other, it is a herd behavior.

Alpacas communicate with body posture. Crouching with tail over back and head low when being submissive, ears back with head up and back and tail held high when aggressive.

When alert to danger they grow what looks like six inches in height standing very tall and alert with ears forward.

Alert alpacas

March

Two females are still head shaking when running to the food bowls, so I have treated their ears by using the dog flea product. Spray some onto a gloved hand and then apply this to the inside and outside of the ears. The ears look a lot less pink on the outside. But the fur has become very dirty where I put cream on the ears.

I had a horrible cold last week, so the alpacas were fed but I only picked up a little dung each time from the females field. The boys are very accurate and always use the same spot, so it is not so essential to clear up after them everyday.

Health Records

I keep a health log for each alpaca on the computer. I note the alpacas name, sex, colour, microchip number and its date of birth. I note down each routine treatment and when the next treatment is due. I write down the date of shearing and when the toenails were trimmed. I record the date when females were mated, the date and sex of cria born. Finally I write down any health issues and the treatment given.

This log gives me a record, at a glance, of my alpaca. The log is useful for herd management, for example which animals to breed and which ones to sell.

Weaning

I normally wean the baby alpacas by separating them from their mums at six to eight months of age. This year I experimented, by leaving the babies with their mums, to see if the alpaca mums would wean their cria.

I noticed one mum was not allowing her cria to feed. If he tried to get his head under her tummy she would kick him away. When she

16

went to the toilet area he would follow her and try to feed, while she was trying to 'go'. She would walk away and then circle back to the toilet area and try again. It would take her several attempts to successfully go the toilet.

It has been a very hard winter. We had one week when the temperature stayed below zero even during the daytime. Although her cria was quite young, he has stayed very healthy. He was already tucking into the hay really well and he had been trying the hard feed from a very young age. I did feed the young ones separately, so the grown-ups did not hog all the feed.

Another mum kept on feeding her cria, although she did make some effort to stop the feeding. When she strained a leg, she was still willing to let the cria feed some of the time. At this stage I removed the cria.

Normally I move the cria out of sight and sound of their mums. This time I experimented by having them in a paddock right next to their mums. This way they could still sleep near their mum but not feed from them. However one cria kept sliding over the fence to get back in with her mum. So I moved the mums into a paddock out of sight. Then I moved another set alpacas to within sight and this seemed to work well! The weaning takes about six weeks. Then the babies can return to their mums. Check young males are not too bothersome, trying to mate with the females. Males are normally moved to bachelor accommodation before ten months of age.

Before weaning the babies get them used to eating alpaca coarse mix and hay. Some babies just naturally copy the adults. Others may need to be fed in a pen next to the adults so they are not bullied away from the food.

Once we are confident they are eating adult rations I separate them for weaning. Making sure they get plenty to eat. Babies do need company when they are weaned. I normally hope to have at least two babies and wean them together. A maiden female can provide company. Another way is to use a gelding for company. I used to have one gelding and I put him in with babies. They were then not worried, but terrified of the gelding. He would chase the babies around and tell them off a lot.

How to separate mums from their much loved babies? Well I use hurdles to make a temporary pen. Make the pen joining onto the field gate. Have the alpacas in the field as the pen is made. Alpacas are so curious they may all walk into the pen. Shut them all in. Put down some feed for the mums open the gate to the other paddock and shoo the babies through.

When babies are being weaned they are super friendly and get right under my feet as I tidy up the paddock. This is an optimum time to halter train.

9th Mar

I decided to try the mite oil again on the troubled ears. They are actually very crusty around the base and on the ear edges. The crusts felt very thick and concrete hard. I put on some baby cream (zinc and caster oil) to see if they soften and come off.

Don't Panic

I had a query, from a friend, about an alpaca that was standing with its mouth open and not eating. The owner thought the animals jaw had locked. However, the animal did recover after about fifteen minutes. Stress does cause alpacas to stand, often quite still, with their mouths open. Bottom lip very relaxed and dribbling.

I first saw this when two very pregnant females, full of hormones, had a disagreement. Afterwards one of them stood very still with her mouth open. This is quite upsetting to see and I was a little worried, however, I was sure it was due to stress. Sure enough she did get over it and they all went back to normal.

10th Mar

Penny is lying down on her side following eating. This happened in the evening and after her breakfast. I called the vet and he came straight away. He treated her with an anti-inflammatory. He suggested reducing the amount of coarse mix she gets and feeding less, more frequently. He said if she had a high worm burden this could aggravate her condition. He was really telling me he thought it was indigestion. He also said it could be gastric ulcers which is incredibly bad news.

Getting Alpacas into the Catch Pen

How to get alpacas into the catch pen? I think the best way is to feed the animals in the catch pen every day. When I first had alpacas I got them used to going into the catch pen. Then we practiced putting on the head-collars and holding the alpaca around the neck. When the vet or shearer visited it was reasonably easy and stress free. This is especially important for nervous animals. They need to settle into their new home and to learn to trust their new owner. When it is time to trim toenails or give vaccinations the animal will be more relaxed in an enclosed space and will allow itself to be handled.

Eating in the catch pen

It seems a bit obvious but I don't give the animals a full feed if later in the day the alpacas need to be coaxed into the catch pen. A little hungry and they are more likely to go in the catch pen even if it is not the normal feeding time.

If the vet is coming it is better to put the animals in the pen before they see the vet. I ask the vet to wait out of sight for a minute while getting the animals in.

April

Today I bought more dog flea spray; the last lot was very old and has all been used. I used this to treat the alpacas that have been shaking their heads. Spray onto hand and then rub around the outside of the ear and the inside also. I was very careful to make sure I had not missed any areas. I also treated the surrounding head area in the same way as the mites do move around a bit.

Weaning onto Lush Grass

We moved a pair of cria onto a new paddock to wean them. The paddock had very nice green lush grass and the eight month female developed a dirty tail. She seemed to be in very good health otherwise. I spoke to other breeders for advice and they suggested keeping her in a pen with a concrete floor and feeding her hay. I tried this for three days and she still had a dirty tail, but was otherwise bright and perky.

At this stage I called the vet. He recommended some medication and to keep on with feeding hay. She continued to be in very good spirits and very slowly the stools firmed up.

10th Apr

I am still putting out hay for the alpacas. The spring grass is now growing. However the alpacas would rather eat grass than hay even though there is only a little of it. So for the females I am putting out a slice of hay every few days. Last summers babies are being weaned at the moment so I just put them out a slice every now and then. They are living on a field with good grass and they are getting a good portion of alfalfa and coarse mix.

14ᵗʰ Apr

Well I have noticed that both my head bangers, alpacas with ear mites, are now very serene, with no head shaking. The dog flea product seems to have done the job. I wonder if my last tub was just too old to be effective. Or maybe mites were living deep down under the hard ear crusts and this medicine could not treat them. Perhaps in the future I need to get the crusts off and then treat for mites. I think I will just use the zinc and caster oil cream to soften crusts and see if I can ease them off. It is bliss seeing them all look so well and happy.

Poisonous Plants

Check paddocks and hedges for plants and trees that are poisonous to alpacas. Alpacas have evolved in South America and may not recognise poisonous plants that grow in the UK, or other countries.

The British Alpaca Society has a list of poisonous plants on their website. When checking the land I remember that alpacas have a very long reach and will even 'stand-up' on their rear legs to get to something they think is desirable. We check paddocks regularly as some plants, such as ragwort, arrive as seeds blown on the wind.

19ᵗʰ Apr

Yesterday we vaccinated the alpacas for bluetongue disease. Fortunately this vaccine is reducing in price. When the vaccine first came three years ago it was £20 for 20ml or £50 for 50ml. It is used at a rate of 1ml per animal so £1 a go, but the bottle must be discarded 8 hours after opening. For only a couple of alpacas to do it is a little expensive. Now it comes in 10ml bottle and the price has come down a bit too.

The alpacas did not want to go into their pens. Did they remember our friend from the last time he kindly vaccinated for Clostridial disease? Or was it his children running around with my child that was causing the problem? Anyway eventually they were all coaxed into their pens and vaccinated.

Today the lady alpacas are not very hungry. My greediest alpaca did not want to eat. She seemed very reluctant to put her head down and dribbled at bit. This is a worry; she has two months to go before her cria is due. I will monitor through the day and see how she gets on. Later in the day she was eating as normal, or should I say, greedy as normal!

I talked to other breeders about reaction to the Bluetongue vaccine and they have experienced alpacas who just wanted to lay down, not even getting up to eat yummy grain. They saw reactions to the Bluetongue vaccination a week after application with 'severe flue' type symptoms, but without the sneezing. With time all the alpacas recovered back to normal.

Feeding Only Grass

Some alpaca farms try to feed just grass and hay to their alpacas. They keep paddocks with rich grazing for pregnant mums and young stock and protein poor paddocks for geldings, resting males and resting females. The land is treated with manure and reseeded, at the beginning of the enterprise. Specialist seed merchants can mix a blend of seeds suitable for alpaca pastures.

Soil can be tested for ph levels and minerals, when these are correct the grass grows well and will help to keep the alpacas healthy. Adding manmade fertiliser will make the grass grow but may take the minerals out of balance.

First Cria

The first alpaca cria born on my smallholding arrived three weeks before the date I thought it would be born! The grass had grown well that summer and the alpacas were having alfalfa as a supplement as our breeder suggested. The alpacas loved the grass and were not too keen on the alfalfa.

My daughter and I went out at about eleven this morning to put out some feed and check on animals. I thought 'what is that in the grass?' Shiny and pink, on closer inspection it was a placenta. I looked around the paddock and there was a gorgeous cria sitting up and 'smiling' at me.

I panicked of course, worried about the cria being premature. I went a grabbed my scales. I had a sling to put the cria in and a spring balance to read the weight. The cria wriggled a lot, it was tricky to get into the sling and then the spring balance went up and down as the cria moved. So I just about got a meaningful measurement. Cria weighed over 6kg, a perfectly normal weight so not premature, phew!

Babies can weigh between 4 and 10kg. They grow and put on weight really quickly in the first few weeks. I weigh daily for the first week and I then weighed the baby once a week for a month. Finally, I just checked the cria is getting bigger and the back feels flat. This would be obese in an adult but perfectly normal in a cria.

May

This morning we started shearing at just before nine o'clock. And we soon stripped-off our own fleeces, as a little heat wave hit our paddock. We had shorn eight animals by eleven thirty. I think the alpacas were very pleased. The previous week the females were almost panting. Now with the fleece off they must be feeling much more comfortable.

Shearing needs two people

10th May

Alpacas do eat a lot more after they are sheared, I can tell because I seem to be collecting twice as much dung!

Have a Cria Box

It is good to have a 'cria box' containing everything needed for birthing babies. My first alpaca was born a lot earlier than I had expected and I felt quite panicky but all I had to do was grab the box which contained all I needed.

First cria

15th May

I have been told that alpacas prefer to feed from a trough. So I am trying feeding using a trough. The alpacas are pushing each other with a lot of spitting and this is even when they have been on good grass all day long. My best feeding strategy is feeding bowls, separated by three meters in distance and to put down one more bowl than I have alpacas. I have modified this now by replacing two bowls with two troughs and this is working very well.

Penny has developed a lovely full udder and because of this sign I expect her cria some time in the next three weeks.

Buying Hay

How much hay should I buy? It is best to buy many bales, as we don't know when snow will fall. When the grass is gone and the weather is cold the alpacas need to eat lots of hay to keep warm. A good time to buy is when the hay has just been cut in early summer. The price is low and the hay can be stored to keep it in the best condition. I put the bales on wooden pallets so the air can circulate under the bales and leave a space between the bales and the walls. Pack the bales so air can circulate because new hay still has a high moisture content. Finally keep the hay in the dark and make sure rain cannot reach it.

I normally budget for at least one bale per alpaca per winter month. Some years hay is difficult to source locally. We have just had two very dry springs with almost no grass growth. Talking to my neighbour, they usually cut about one hundred bales of hay. Last year it was very dry in April and May and they only got twenty six bales from the hay field. This year it has been so dry they think they will not manage to cut any hay as the grass is only three inches long! This means there may be very little hay available to buy locally. The larger feed merchants should be able to buy in hay from other parts of the country but this will put the prices up.

18th May

The two yearlings Alice and Bramble have had their Bluetongue booster today. I vaccinated without an assistant. I stood the alpaca parallel to the fence in the corner. The corner stops the alpaca going backwards, my knee stops them going forward and my body keeps them against the fence. I then lean over the neck and vaccinate.

Alpacas don't seem to react much to the needle. The young ones are far more frightened of being restrained by the neck.

21st May

The sheared alpacas are very happy during the daytime, but at night the temperature is still going down to freezing. At night the alpacas are on the sheltered stable paddock. During the daytime, I move the alpacas to the orchard paddock that has lots of grass.

Now moving alpacas to the stable paddock is easy because this is when the alpacas get their daily ration of alfalfa and yummy alpaca mix. But this morning, I had a little alpaca rebellion. To get to the orchard paddock the alpacas have to run through the corner of the old menarge. Now the menarge has longer grass than the orchard. Yes, you can see what the alpacas are thinking, and yes, they did run off. However, if I rope off the corner as a corridor, they are very obliging.

Looking daily now to see if the alpaca cria is about to come. I think the cria will arrive some time during the next three weeks.

How Many Paddocks?

One thing I had not considered when I started breeding alpacas was how many paddocks I would need. I started with a paddock for my pregnant females. Then I needed a paddock for weaning the babies. After this I needed a paddock for the boys, females go in with their mums. Finally I found I could not keep entire males together. They need to be kept where they cannot see or smell the females or they will fight. My smallholding was too small to achieve this, so I needed adult male accommodation, one paddock for each male.

Also young males will probably be bullied by adult males. So another paddock is needed for males who have not reached breeding age. Of course a gelded male can go and live with the females.

Adult male alpaca

30th May

Penny's cria was born today. At feeding time there were no signs of labour. Later on we were outside playing with our dog. Our daughter spotted that the cria was making an appearance.

The cria arrives

Lisa has not seen an alpaca cria arriving before and she rushed inside for her camera. The head and front feet had appeared.

Dry the cria and spray the cord

We waited for the neck and the whole of the front legs to come out. Mum stood up and sat down several times at this stage. Pushing the shoulders through takes the longest time as this is the largest part of the cria.

The mum and the cria were not distressed, and eventually the cria popped out. My daughter passed me the towel to dry the cria off and the iodine to treat the navel cord. It is very important to make sure the cord gets treated to stop pathogens from entering into the cria's body. The whole herd gathers round to greet the cria.

The herd greets the cria

Normally the placenta comes out within an hour. However this time it took three hours. Mum was not distressed in any way. She was eating and looked very happy. Once the placenta was delivered the cria fed, immediately, for the first time. I then felt very happy. The mum will often not let the cria feed until the placenta is delivered. The next thing for me to do is to check the placenta is complete, check for tears or missing bits. If there was a problem with the placenta, it could mean the mum has retained some of the placenta and this would require treatment by the vet. This time it all looks fine so I dig a small hole and bury it.

Tonight and for the next few nights I will put a dog coat on the cria, to keep it warm. If there was heavy rain I would put the mum and cria in the stable. I would probably leave the light on as babies are programmed to look for milk in dark places and tend to spend ages trying to feed off the wall if the light is out, this is sometimes called 'a wall baby'.

Guarding the new cria

The picture shows the herd guarding the new cria as it rests. Other adorable things that alpacas do with new babies: all rush and go to the toilet when the new cria investigates the 'poo pile'. Gently spit at the cria once it starts to walk around as if to say I am not your mummy!

Also when the alpaca has just been born the herd is silent and only the cria and mum talk. In this way the cria gets to know which alpaca is its mum. Mum and cria alpacas kiss and we think it is another way of passing anti-bodies.

Mating

Which male should I choose to mate to my females and how does the travelling for getting them together work?

The best stud males have highly desirable conformation, fleece characteristics and have proved successful in passing on their genes to their offspring. Consequently these males are very expensive to buy.

Owners of female alpacas can purchase the services of this type of male, by taking their female to the male, if they live close by. Or the male can visit the farm and he can travel a large distance. A third option is for the female to stay at the stud farm. This usually means a visit of two months or more as the female should not undertake a long journey when newly pregnant as this stress may cause the pregnancy to fail.

The price of a mating may include two months livery. Or, for a 'drive-by' mating the fee may only cover three visits. Some farms guarantee a live cria on the ground for twenty four hours.

If the female does not get pregnant the mating fee may be transferred to another female the following year. It is probably best to ask for a written contract so there is no ambiguity in what the stud fee covers.

Medicine and Alpacas

At the moment in the UK there is no medicine licensed for use with alpacas. Under the advice of our vet we use sheep and cattle products for vaccinations and worming. For mites we use dog products for topical treatment.

June

We are halter training last years babies. Normally we do this at the weaning stage, six to eight months of age, but I didn't manage that this year. Alice is now a year old and Bramble is nearly ten months old. Alice is accepting the head collar okay, which is good because her mum dislikes having her head touched. Bramble is a very amenable and a quick learner; I gave his head a rub between the ears and then put the halter straight on. He did not even move his head. The next day I did the same and then clipped a rope on to the halter and let him walk freely round a pen with the females in it.

Next time we had his mum waiting in the paddock on the halter led by my husband. We put Bramble on the halter and let him walk freely to his mum. Duncan lead Petra round the paddock and Bramble followed beautifully on the halter.

Worming

When should alpacas be treated for worms? Standard practice was to worm in the spring and autumn. More recent studies suggest the danger time for worms is June onwards, until the weather gets colder in the winter. So this is the time I test and treat for worms.

One vet did advise me to worm every six weeks like we would for a horse. Worming medicines are a type of poison; they kill that small 'animal' the worm but not our large animal the alpaca. We want to expose our alpacas to the smallest amount of poison necessary. Worms can build-up resistance to drugs, so it maybe better to test for the presence of worms by having dung samples analysed. And then treat the alpaca only when we know it has worms.

Sunbathing

Alpacas like to sunbathe and sometimes the whole herd lies down together. They like to get the sun on their stomach. To do this they lie down on their side and then pull their head back and legs up.

Enjoying the sunshine

Handling Cria

For the first six months we let the cria learn all about being an alpaca. We don't interfere by handling them a lot. Then at weaning stage we handle them more and halter train.

We have heard stories of bottle fed alpacas being over handled and then treating humans like alpacas. This becomes a problem as male alpacas do fight to find out who is the alpha male.

9ᵗʰ Jun

Cria is now one week and two days old. She now looks huge! She is spending much less time asleep, and when she is resting she holds her head up straight instead of her neck looking like the stem of a droopy tulip.

She is playing with last years babies who are now coming up to a year old. She has started to try a graze with the rest of the herd. This is good and bad because she doesn't know what she is doing yet and will try and nibble on anything. On the whole she looks so much more robust and in charge of her legs than she did last week.

Mating Records

I like to keep a 'mating book'. Here I write down all the details of the mating and births from each female alpaca. I record the date and time of the mating and what the weather was like. If the weather is very hot, the males can become temporarily infertile. I record the location of the mating. It could be a pen mating: in the male's paddock or in the female's paddock. Mating could be in a special mating pen in neutral territory or on another farm.

I also record when the female started to refuse the male and was thought to be pregnant. This is called spitting-off as the female may spit at the male, instead of lying down and letting him mate. Finally I record the date of the cria arriving and timings of each stage, if I can.

Mating records are great to look back on. They help me to know if a female always gives birth two weeks early or if she needs several attempts to get pregnant. All this knowledge helps with the farm management.

Water

Alpacas require access to clean water, at all times. In the summer, my alpacas like to paddle in the water trough to keep cool. This means changing the water daily. I also provide them with a water sprinkler when it is very hot.

Drinking from the hose pipe

My alpacas love to use the water sprinkler to get cool and to have a drink. When I turn the hose on they queue for their turn to enjoy a shower.

July

The weather has been extremely hot and dry. I have been putting the spray attachment on the hose and tying this to the fence. The alpacas have been enjoying a shower. Alice likes to drink from the spray attachment.

22nd Jul

The grass is not growing due to heat and no rain, so I am feeding this year's yummy hay. I am also feeding half a scoop of alfalfa and half a scoop of alpaca mix plus one hand full of soaked sugar beet, three times a day. This is between five adult alpacas.

Nina eating from the trough

38

The alpaca cria born eight weeks ago is eating hard feed out of the feed trough with the other alpacas. We have named her Nina.

I put out as much feed as they can eat in five minutes. I have experimented with putting out a generous amount of alfalfa when I have a female feeding her cria. However, I find that a lot of the alfalfa is wasted. The stalks are usually discarded on the ground when they stand in the feed bowl.

With the current feeding regime all the food is eaten even the stalks. Also it is good to feed the pregnant alpaca a little and often, as the room she has for food in her stomach decreases.

First Aid

As well as a cria box, I keep a very basic first aid box. In this I have wound powder to put on small cuts. Foot rot spray for trouble between the alpaca toes, such as old mite damage that may become inflamed. I also keep a product in case of fly strike. This is rare but can happen, for example if faeces build up on pieces of bramble stuck in the alpaca fleece. I keep bandages, dressings and 'sticky bandage', although fortunately I have never used these. I also keep fly repellent spray and mite spray.

29th Jul

I now have eight white alpacas that are for sale. Several people come to look who are really very keen to buy alpacas. Each animal has a data sheet, with fleece statistics, prizes won and pedigree. One couple came and looked for two hours and could not decide on which two of the eight they wanted to buy.

Another person came twice and was still undecided. Both sets of purchasers took away the details and rang through later and made the purchase having read the alpaca information sheet.

Mating with an Inexperienced Male

Chester was approaching three years of age, time for him to start his career as a stud male. We put him in a pen with a female who had a two week old cria. Chester stood and shook! We tried him in the pen with a female everyday and still nothing happened.

Mating alpacas

After about one week of trying him in the pen he gathered up all his courage and had a go. When the female sat down for him he was delighted.

He needed a few goes to think about how mating works. Sometimes he just wanted to play with neck wrestling and a bit of friendly biting! Eventually he worked with enthusiasm and we had two lovely cria from him the next year.

August

I look for pink and very swollen nipples, this can be a sign of labour. Also look for consistently walking around with the tail held high away from the body. Petra has lots of milk today and she went into labour this afternoon at about five thirty. She has just taken endless trips to the dung pile. Then she stopped eating and is just standing away from the others by herself. She has also been rolling frequently.

I am concerned about her so I watch what is going on. I go outside and give the alpacas supper at about nine o'clock. It is still light at this time of year. Petra comes over and tucks into her food. I check again at midnight and she is lying down with the others, nibbling at the hay. She seems quite comfortable.

I check again at five o'clock in the morning and Petra is lying down with the others and nibbling on hay. I give the alpacas breakfast at nine o'clock and Petra happily walks over and eats with gusto.

At nine thirty labour starts again. Petra keeps visiting the dung pile and then she moves away from the pile and starts to push. Alternately pushing, lying down and occasionally rolling. In fifteen minutes the nose appears, followed by the head and front legs. The cria then hangs down and fluid drains, this bit always seems to take forever. Petra is in no distress at all and not pushing much. The cria is okay too, occasionally moving its head because of a tickling fly. Sometimes the babies complain at this stage and can be quite noisy. The last stage is very quick the cria slips out once the shoulders have been delivered.

I rub the cria down, remove the fine membrane and spray the cord. It is a fawn male baby. All the alpacas gather round to greet the

41

new cria. The cria makes lots of gasping noises as he breathes; he seems to be moving a lot so I don't panic.

The cria rolls around and gets into an upright position within a few minutes and then stands quite quickly. The herd again gathers around to sniff and nudge him along. Breathing is all quite normal now. At this stage I put him in a little pen with mum, so they get to know each other. Petra passes the placenta about one hour later. Once this process starts it is all finished within minutes. Passing the placenta stimulates the cria. He stands and passes his first stools; this meconium comes out in three little lumps. Petra turns and sniffs his bottom and eats the last lump of meconium he passes. Does this contain some elements that she needs?

Three hours pass by and he hasn't managed to feed. So I make up a bottle using powdered colostrum for alpacas and give this to the cria. The powdered colostrum gives him a boost in energy to keep trying to feed. I always joke that it tastes so bad the babies jump straight to their feet and try to feed from their mum. Anyway this did the trick and he was up and trying to feed from Mum, immediately. I notice the tendons in his back legs make a clicking noise as he walks, I think this can be a sign of prematurity.

Newborn Cria and Feeding

Newborn cria have very long legs, they tend to be a bit unsteady and topple over. My friends call this the 'Bambi' stage. Also they are programmed to look in dark places for milk and they try under the tail and between the front legs. They get to the right place and then topple over.

Sometimes it takes hours to get the hang of drinking milk. We mix powdered colostrum with warm water and feed this from a bottle if we think the cria had not fed in the first few hours. It gives them a boost and they keep on trying to feed from mum until they are successful.

Mating, Try More Than One Female

I hired in a stud male. We put him in the pen with our female and nothing happened.

Normally the male jumps on the females back and she goes into the sitting position with all four legs tucked under her. Sometimes she will roll onto her side during the mating.

Mating alpacas

So we had a cup of tea and some cake! Still nothing happened. We brought over some of our own boys as competition and still nothing. In the end we tried another female and he started mating immediately!

5th Aug

Little male alpaca is now named Fawn. I have been keeping him in a paddock with just his mum and an auntie, while he gets to know his mum and learns to feed. The second day I tried putting the main female herd in with him. The little female cria, Nina who is nine weeks old, wanted to play with him. Fawn was a bit frightened and kept running away. Then he would put his head and neck through the sheep netting, along the fence and not know how to get himself out. So I took our female herd away again for twenty four hours.

6th Aug

Today Fawn is in with the rest of the herd and doing well. He has had a little run around and gone to say hello to all the adults. They have told him off in the nicest way, as if to say I am not your mummy. Later on Nina noticed him again and they had a little run around and then Fawn went back next to his mum for protection.

We also had a little excitement today. An alpaca gave a great big alarm call and all the others started to run around. I went to have a look at what was bothering them and a herd of eight cows came running across the adjacent field of sugar beet to look at the alpacas. We don't know where they escaped from and they moved away again as quickly as they arrived. By the time we had our shoes on to go and investigate they had disappeared from view.

7th Aug

Yesterday there were lots of flies in the air. I sprayed Petra's cria with repellent in the morning and tried to spray the rest of the females as they had their breakfast. Alpacas are not too keen on fly spray. They are suspicious of the canister and of the hissing noise it can make. During the day Petra had some discharge from her rear end. This is normal after having a cria. Discharge looks pale pink and is like jelly in consistency. Sometimes there is a blob of it on the ground after the alpacas have rolled. Petra must have swished her tail and the wrong moment and she has a damp rear. The flies loved this and were all on this area. I gave her a feed so I could treat her with the fly spray and prevent fly strike.

Nina is very gentle with little Fawn. She tries to get him to play run around but she doesn't jump on him. This is good because he would be flattened by her.

11th Aug

The discharge from Petra has 'dried up' completely, after having her cria. She has plenty of milk and is eating really well.

Baby Fawn looks much more grown-up now. When he is lying down he holds his head up really well. When they are very young their head is almost too heavy and droops backwards. When he was newborn his hocks were bending out and very angled. The back legs are now straight in line with the body as they should be.

He also had very clicky tendons when he was newborn, and this clicking is now less noticeable. In the first week the babies drink milk, have a little run around and then flop down. Now Fawn is a week old he is trying to imitate the adults and standing to graze. Before he would flop down anywhere and now he is following his mum. When it started to rain she led him over to a tree with a great big canopy and they both sheltered from the rain beneath the tree, so sensible. When the larger cria wants to play Fawn knows he can run to his mum for protection if he feels worried.

12th Aug

Always prepare. I had a customer come to buy some alpacas. I wanted my stud male to walk into his catch pen so I could put him on a halter and show him off. He didn't want to be moved away from the female alpacas in the field next door so he refused to go in the pen. I tried for about 10 minutes in front of the customer. In the end a caught him in a corner and put him on the halter.

The same customer returned for another look yesterday. This time I prepared, by putting the male in his catch pen before their arrival. He was impeccably behaved as I put on his halter. My customer was impressed by how easy he was to handle.

Ready for sale

25ᵗʰ Aug

Mating day, it is over two weeks since baby Fawn was born. I have wormed Petra and she is looking at the stud males everyday. We put her in with Chester and she sits down for him straight away. Petra normally gets pregnant on the first sitting. In a few days I will need to check she has taken and check in a few weeks that she has not absorbed the pregnancy.

Rotation of Paddocks

Rotation of paddocks by moving alpacas onto new pasture every three weeks, and daily removal of droppings, helps to keep alpacas free from worms. This year I had rather too many animals and rotating paddocks became difficult.

I had extra paddocks for quarantine of new animals and a visiting stud male.

Since the summer I have sold a stud male and the visiting male has gone home. The new stock has completed quarantine and joined our other alpacas. I have sold the young males. At the moment I have two paddocks in use, so rotating paddocks becomes easy again.

When the grass stops growing in the winter, I stop rotating paddocks. The pregnant females would rather eat grass than hay in the winter. Growing babies requires good nutrition and hay is better than winter grass. So I feed hay and keep the females on meadows with short grass and shelter.

When the grass starts growing again, with longer hours of daylight and temperatures above ten degrees, I will start to rotate paddocks.

Maiden Female

A two year old female, we waited all summer for her to be ready to mate. She liked the male but would not lie down for him. We let her watch the other girls mating and she still refused to lie down. We decided to take off the head collar and move away from the mating pen. This worked and she lay down.

26th Aug

Fawn is almost three weeks old. He seems to spend his time either playing or feeding. The days of resting between feeds have ended. This summer we had two babies out of the same black male. One baby has a very fine fleece that is not dense. And this baby fawn has a very dense fleece, so I worry much less about him on rainy days.

I have noticed that Fawn has copied the other cria and is now nibbling at the hay and trying to eat the hard feed. He will soon be old enough to have his clostridial injection.

Where to Mate

Our stud male went to work at another alpaca farm. He had two females to mate and he was put in a pen next to all the females. He then found it very hard to concentrate on the girl in the pen with him, when there were lots of other females outside the pen.

We returned seven days later to do spit-offs and this time we put him in the mating pen together with the two females and moved all the other females to a paddock about fifty meters away. Our boy concentrated on the two females in the pen and got the job done!

Fleece

Why is alpaca fibre so comfortable and warm to wear and yet so light in weight? Well alpaca fibre looks a bit like hair under a microscope. It has a cellular composition, which helps to make it lighter and warmer than sheep's wool. Alpaca wool is second only to silk for strength, has less scales than sheep wool and no lanolin. The individual strands of alpaca fibre are finer in diameter than sheep's wool.

A sample of fibre from an alpaca normally measures between 16 to 30 microns (a micron is one-thousandth of a millimeter). A good adult alpaca fleece would measure 25 microns or less. For comparison human hair measures between 90 to 100 microns, and cashmere would measure between 11 to 18 microns.

Some farms have details of the alpaca fleece statistics on their for sale adverts. There are laboratories that measure the quality of alpaca fibre. The results are shown in a fibre histogram. The histogram represents the quality of the fleece at the time of sampling. The fibre produced by an individual alpaca will change with differences in diet, pregnancy and age.

The fibre histogram shows a graph of the fibre measurements, with numbers for AFD, SD and CV. These terms in the histogram are:

Average Fibre Diameter or AFD is the fibre diameter in microns.

Standard Deviation or SD is the amount of variation in a sample from average or AFD. Five percent would be good but this figure ideally would be as small as possible.

Coefficient of Variation or CV is the SD/AFD x 100 and twenty percent would be good.

The percentage of fibre greater than 30 microns is also documented. This is known as the 'Prickle' factor that is the scratchiness in clothes associated with coarser fibres over 30 microns.

49

28th Aug

The vet came to insert micro-chips for the alpacas that have been sold. Once the chips have been inserted the animals can be registered with the BAS. I bought micro-chips that comply with the BAS requirements.

Our new female is registered with the BAS but she had not had her micro-chip inserted. When her breeder went to put it in, it went a little wrong and the chip fell on the ground and was discarded. So we needed to put a new chip in today. I was a little worried because the new female doesn't know us or where the catch pen is. With a little gentle encouragement she went into the catch pen and she let me put a halter onto her. I think she has only worn one a couple of times before. Her breeder said she knows about wearing a halter but not how to walk on the halter. She is a very quiet girl and stood like an angel while the chip was inserted.

The two yearlings were also very easy to micro-chip. They are both halter-trained and stood beautifully. The alpaca that proved a challenge was our three month old. She has been handled but she wanted to jump around a lot and the micro-chip ended up on the ground. She did become quiet when a young male put his head through the fence and sniffed round her top-not. Then the vet managed to get the chip in. It is lovely when the young alpacas 'look after' each other.

Showing Alpacas

This is fun and the other competitors are really friendly. There is the chance to see really good alpacas close up and ask lots of questions about how things are run on other alpaca farms. The BAS, the British Alpaca Society, arranges all the entries, takes the money and allocates pen space, generally ensuring that the show is run in an orderly and timely manner and in accordance with the BAS rules.

Grassroots is their online system – it is used for animal entries to the shows. Each alpaca should have Public Liability Insurance in place.

Chester wins a prize

At the show the alpacas are kept in pens, a little like sheep pens, often in a large barn or marquee. The judging area is similar in size to a Donkey ring. Prepare in plenty of time for the showing season by halter training show animals. Animals need to be able to walk on the halter and to stand still for the judge to examine them. It is good to get the animals used to walking past parked cars, because stock transport is often allocated parking away from the alpaca pens.

Remember to take water buckets and hay, alfalfa and feed bowls. Alpacas are bedded down on straw in the pens – this maybe supplied by the show or the owners own straw. The alpacas are happiest if penned in small groups. Males and females are penned separately ideally with an empty pen between the males and females.

Some shows are two day events and the animals overnight in the pens. When the general public has left the alpacas can be taken out for a little walk to stretch their legs.

Given that huacaya need to be sheared annually, the fleece should be as long as possible, so shearing is left until after the shows.

27ᵗʰ Aug

Last year our male had blood between his toes, and this area was very crusty to feel. There were a lot of flies around his foot. I treated the foot by washing the area and used wound powder, for two days. The problem was not solved. I called the vet. Our surgery is a teaching practice and a young vet checked our alpaca. She said the foot smelt a bit 'cheesy' which meant it was infected. She gave him a long lasting antibiotic injection and gave me some foot rot spray, which has a topical antibiotic, to apply twice a day. This did the trick nicely and the foot healed. It was interesting that the alpaca was not lame and the foot did not seem to bother him in any way.

Well that was during last summer. This week the foot looked a little red between his toes. And then I noticed a couple of flies crawl between his toes. So I called out the vet.

The vet came and he diagnosed either: rain scald or foot rot or mites living deep under the crusts. He said maybe it didn't heal properly last year. He gave an antibiotic injection which lasts four days. The vet recommended washing out daily with warm antiseptic solution and then using the foot rot spray. He said to scrub the scabs off with a nail brush. The stud male has been a real angel about having his foot treated. He just stands in the corner with a hand on his neck while I wash his foot. I do this by using, a gentle squirt, from a syringe full of antiseptic solution. Then I apply the purple spray.

I haven't been brave enough to do the nail brush scrubbing. When it is healed up I am planning to put on the zinc and castor oil cream to soften the scabs, so they can be removed.

Healthy stud male

September

The apples have started to fall from some of the trees. Every evening I go collecting apples. The rotten ones go in the brown recycle bin and the good ones go to feed the alpacas. They can easily cope with two small apples in the morning and another two in the evening. By picking up all fallen apples we also keep down the number of wasps.

Mating our new black alpaca female with a visiting male, Prince arrived first thing in the morning for a two week stay. We put him in the orchard paddock, with hurdles dividing the paddock into two sections.

In the afternoon our new female Delta and her cria arrived, we unloaded them into the far side of the orchard paddock, separate from Prince and went inside for a cup of tea.

Prince decided he was ready to mate and managed to push the end hurdle open. Delta went into the cush position for him almost immediately and mating began. After about 20 minutes they decided to stand up and we took this opportunity to put Prince back in his own side of the paddock. Once we had done this we made a catch pen, again using hurdles, right where Prince had pushed through. This worked very well and Prince was happy now that he had mated and stayed on his side of the temporary fence for the rest of his working holiday with us.

Over the next fortnight, we did three spit-offs using Prince with Delta. We have halter trained our own stud males. When we do a spit-off with them we have them on a halter. We put the female into the mating pen with the male. If she does not want to mate we use the halter and lead rope to restrain the male. Prince is not halter trained. I had to be brave and hold him by the neck, while Duncan shooed Delta out of the catch pen.

When I went to choose Prince as a visiting stud male I saw him kick his herdsman in a very uncomfortable place. Prince can be a little feisty when he is unhappy about something. But hanging on to him while his lady friend is ushered away caused no problems at all. And while Prince was with us I moved him very gently to where I needed him and he behaved like a perfect gentleman!

Delta spat-off on day five, day ten and day fourteen, so I believe her to be pregnant and time for Prince to go home. I made a catch pen next to the large field gate, which is located next to the long side of our outbuilding. I then put feed and hay into the open pen. I tied a long rope to the pen and left this trailed across the grass. I then herded Prince towards the pen. The side of the outbuilding forms one side of a funnel, I picked up the rope to form the second side of the funnel and gently asked Prince to walk into the pen. He then spotted the feed and walked in happily.

While waiting for Prince to be collected I chatted over the garden fence with our neighbour. Prince's owner drove in and reversed up to the catch pen. We dropped the ramp on the cattle trailer, opened out the guard-gates at either side of the ramp and put hay on the ramp. We opened the field gate at the front of the pen and Prince happily walked up the ramp into the trailer, loading him took less than one minute. My neighbour commented favorably on how stress free this all was.

Toenails and Studs

I have a stud male called Chester and he thinks he is the man. Last year he didn't want to lie down at shearing time. He jumped up and ran away. It took three of us to put him on the ground and hold him while putting on the shearing ropes!

Now it is time to trim his toenails again. I have decided to try and get him to like us more. I have made his catch pen smaller and I stand in there with him everyday for a short while when he is eating. I take him an apple in the afternoon and have a little chat with him. When I did catch him and put the halter on he was a gentleman. For toenail trimming my husband used a very light pressure on the halter line and stood very close to Chester. I ran my hand down his leg very slowly finally lifting the foot. All feet were trimmed very easily with no leaping around by Chester. Hopefully the next shearing will be a lot easier too!

10ᵗʰ Sep

Delta was wormed and treated for mites before she arrived and she has been in quarantine. She is now fit and ready to join the female herd. With a bit of a shuffle round of animals I put Delta into the small paddock that joins onto the female's field.

There was a great deal of interest, the females and cria all rushed over and rubbed noses with Delta and her cria. During the day the younger females and cria kept returning to the fence to make friends. When left on their own Delta kept patrolling the fence wanting to get in with the herd.

At tea time I decided to open the gate and the two sets of alpacas swapped fields. With the gate left open they eventually all gravitated to the same field.

Delta was very assertive with the other alpacas. Any cria near her got spat at and pushed. She made angry faces at the adults and kicked out her back legs. Don't mess with me! By the end of the day all had settled down and she was moving and eating with the rest of the herd. When I went out with a late supper for them all, she came running over with the rest and ate happily out of our trough.

I did notice when she was first in with the herd, although she was distressed by having strange alpacas near her, she got down and rolled in the communal dust patch. She then did this many times over the afternoon. Was she trying to take on the 'scent' of the herd so she would be accepted?

I have two entire working males who live next to the females. Our youngest male Chip reacts in a very submissive way to babies and walks away crouching, or he 'kisses' them very nicely over the fence.

Our second and older male Chester's reaction to babies is normally lovely and he also gently kisses them over the fence. However Delta's cria is black like her mum and Chester reacts in an aggressive way when the cria goes near his bit of the fence. I wonder if he didn't like her colour or if he knew she was not his cria.

So I could relax I decided to move Chester to a paddock away from the mums and babies. I am hoping he will get used to having a black female around and accept her cria next year.

Mites

Penny has ear mites. She is shaking her head a lot. First I look in her ears to check for seeds or bits of vegetation that may have got into her ear. I then very, very gently use my finger in the ear and find brown wax, its best to use latex gloves for doing this! Brown wax is a sign of ear mites.

I treat her ears with some of the mite powder that I use for my dog's ears. It works really well for my dog and gives Penny a lot of relief. I don't think the mites are completely cured but she looks a lot happier. Before she was walking around with her ears clamped down very close to her head. Now she is walking and eating with her ears in a very relaxed position.

With one further treatment the mite in ear problem seems to be under control. Treating ear mites with a pinch of dry powder is very easy.

11th Sep

Petra is a very good mum. She has really top quality babies that grow quickly. But she is quite strict about when they feed. Once they are about two weeks old she won't let them feed on demand. She leads them to a particular area of the field and then they are allowed to feed. The cria soon learns and follows mum to this spot and has a good suckle.

Some days are still very hot. Our lady alpaca's like to stand with front legs in the water trough and use their feet to splash water on to their stomachs. They try to lie down in the trough but it is not big enough. Then they go and roll in the dust patch or on any spare hay on the ground. White alpacas turn into brown alpacas!

Also the babies love to roll in the hay. If I put hay on the ground for the alpacas to eat they want to roll in it immediately which affects its palatability. So we have made a hay feeder. It is like a four sided cage made of wood. The alpacas put their heads through or over its top to eat the hay. Six or more can easily eat at the same time. It stops the rolling problem and helps to stop the wind blowing the hay away.

Checking for Pregnancy

We are doing a check to see if our females are pregnant. After the first mating we wait five to seven days and then we try to mate again. If the female is pregnant she will refuse to mate with the male, which is called a spit-off. We check her with the male again in another seven days. If she refuses again we then try in fourteen days.

At this stage, I keep checking her reaction to the male, every fourteen days. I check by leading her up to the males pen and see if she has any interest in him. When we reach two months, without interest, I believe her to be pregnant. Then I wait, very patiently, for eleven to twelve months for the happy arrival of the cria. Once the female is two months pregnant it is possible to scan her to check the pregnancy or do a blood test.

When doing a spit-off, females may run round like mad and desperately try to escape from the male. Some will spit at the male and kick with their back legs before even getting into the mating pen. One girl even used to hide behind me so the male could not see her. Some females will let the male jump on their back and then they turn their head and spit at him. In other words the reaction varies a lot but they are always not keen. If a female wants to mate she usually lies down in the first twenty seconds of being introduced to the male. I have one female who flirts with the boys but when I put her into the mating pen she rushes round like a mad thing, trying to escape. Talking to other alpaca owners this reaction is quite common.

26th Sep

Fawn is now eight weeks old. He is eating hard feed from the trough, with the other cria and alpaca adults. He grazes and eats hay from the hay rack.

All the alpacas are scratching a lot and Juanita is probably the itching the most. I watched her and she scratches very regularly which must be exhausting for her. I went and bought more flea and mite spray, sold as a treatment for cats and dogs. I will apply this to all the alpacas. There is also an injection, our vet recommends as the best cure for mites and it does worms as well. However, I am trying not to give injections in the first three months and final two months of pregnancy, so I will give the topical treatment.

27th Sep

All the alpacas come in for breakfast in the catch pen. I set the flea spray on the finest mist. I spray Juanita between her toes, under her belly down from the top insides of her legs, round her udder then along her back. She is happy with me doing this. She is eating so happily occupied. The spray doesn't hiss when set to fine mist. After eating I notice she has instant relief from all that itching. Looking carefully at her she does have little red bumps down the inside of her legs, which look to me like a reaction to mites.

October

I used to just feed alfalfa with seaweed meal as a food supplement. However, some alpacas got very thin, over the winter as they were still feeding their babies. So I started to feed 'alpaca mix' available from our local mill. The local mill makes a racing camel mix, which they export to Saudi and as a side line they also make an alpaca mix! This mix works well, keeping the 'weight on' my alpacas.

The alpaca mix is specially made with all the things an alpaca needs. So what to do with the three bags of seaweed meal that I used to feed with the alfalfa? I spread it over my stable paddock. I decided to treat this paddock because it is where the alpacas spend the majority of the winter, while the rest of the grassland rests.

Having spread the meal I really forgot all about it. Looking at the paddock today, the grass is bright green. There are lots of worm casts about. It is a very fertile and healthy paddock. As seaweed meal is so good for alpacas, I have decided to put seaweed fertilizer on all my paddocks.

Prepare for Sale

Prepare for sale and don't leave 'halter training' too late, practice early. Have all medicine for pre-sale treatments ready. Have halters and food in stock to supply with the animals. Get those toenails trimmed and keep on top of unsightly skin mites.

7th Oct

After deliberation I gave Penny an oral wormer. She was last wormed in January and March with an injection. As worms may build up resistance to a type of drug, I give her an oral wormer, with a different active ingredient. I also wormed Bramble by stealth. I put his wormer in with his feed and he happily gobbled it down.

8th Oct

Our ten month old alpaca Bramble has been sold. We took Bramble to a friend's alpaca farm. From here he will be collected, along with another alpaca, and go to his new home.

Before leaving, I shuffled some alpacas into a different paddock, so that I could walk Bramble through to the loading pen. Then I moved two hurdles to make wings that 'attach' to the cattle trailer and to the loading pen. Then we reversed the trailer into position and kept the ramp in the up-position. We walked Bramble from his field to the loading pen.

He watched with interest as we lowered the ramp, covered the ramp in hay and connected the wings to the catch pen and trailer. All this was done quietly and carefully. Bramble became more and more curious. He then walked happily up the ramp into the trailer and this is his first time. It goes to show that being quiet, calm and gentle pays off when handling alpacas.

How Black Is Black

I look at the very woolly alpacas in our field and I ask myself how black is black?

Looking at some animals for sale on the internet, with full fleece, the black has faded to dark brown on the tips of the fleece exposed to the sun. I have been looking at my black girl who is in full fleece. She is the colour of coal, a blue black and glorious to look at, not a trace of brown anywhere!

9th Oct

Today we visited a farm that purchased two of our males to use for stud work. The boys have been living there now for a couple of years.

Our friends were preparing to move one of their young males ready to go off to a new home. This male had been trained to wear the halter and one of our old boys was being walked in front to lead the way. Unfortunately the young lad didn't want to play that game and he had to be pushed along. I was so pleased to see my old boy walked absolutely beautifully. I remember halter training him. It took about five daily sessions that lasted five minutes.

Generally I start to halter train the babies at about six months of age. I feed mum and youngster regularly in the catch pen. So they become used to being near me in a confined space. Then I move on to rubbing the head between the ears and around the face, so they are used to having their face touched. Then I put the halter on and I follow them around the feeding pen attached to them by the lead line. On the next day I ask them to move by pulling gently and then releasing the halter line and all they have to achieve is a couple of correct leg steps. Finally, they have five sessions, following a halter trained alpaca on a walk of no more than a few minutes. The alpaca can be their mum and the last sessions can be with another newly trained alpaca. Now this is really all I did with this old boy before delivering him and he walks like an absolute gentleman.

15th Oct

The females are grazing by the house on nice long, green, grass. The temperature is in the teens, and the grass is still growing well and has a good nutritional content. Chester normally lives on the orchard, next to the house-grass. When the females are let out to eat this house-grass, I move Chester on to the small paddock by the old menarge. I move Chester so he is not tempted to kiss the females over the flimsy orchard fence, which could collapse.

I would like to put the females onto the orchard paddock. But I need to fix the fence first. It needs an extra line of wire because I have one very clever girl who puts her head through the sheep netting and pushes as hard as she can to reach the grass on the other side. Eventually this fence will break so I must add a line of wire to stop her putting her head through.

17th Oct

We have sold four alpacas and loaded them into our cattle trailer to go to their new home. To load them the females are led on halters into a catch pen. The trailer is already parked just in front of the pen. We let down the ramp and put down the guard-gates at the sides of the ramp. Then we used cattle hurdles to fill in the gap between the guard-gates and the gate to the catch pen. We opened gate to the catch pen and the first lady was easy to lead onto the trailer. She was pleased to go in and eat out of the food bucket and her cria walked in with her. My husband stayed on the trailer holding her lead rope. I passed him the end of the lead rope belonging to the second female. She was frightened of loading so I lifted her a little under the sternum so that her front feet were on the ramp. This gave her a lot of confidence and she was happy to walk up the ramp. We took off the head-collars and left them to settle in.

Next to go on was the stud male. We travel adult males and females separately. So the partition was moved across to shut the females in the back of the trailer. He was an absolute gentleman and walked up the ramp beautifully. I think he really wanted to go in and see the females. He had been watching and knew exactly where they were. Before loading the animals we had put hay in the trailer for them to eat and we had put hay on the ramp. If the ramp is disguised the alpacas are more happy to walk on in.

In the UK, remember to take the self declaration certificate and the transport authorisation certificate. Contact Defra, well in advance, for the current rules on animal transportation. Take enough head-collars and lead-ropes for each animal, just in case they need to be unloaded before reaching the destination. Take a bucket for water as well just in case. If it is a long journey the animals need water and hay to eat and regular stops are needed for the alpacas to go to the toilet while the trailer is stationary and for cria to feed from mum.

Also take a map of the destination. Take phone numbers, breakdown insurance, mobile phone and wallet.

You know the address and the postcode. Ask where the field is. Does the field have vehicular access? Ask if they want a phone call on arrival. All these checks can save confusion. I have delivered non-halter trained animals for fellow alpaca farmers, and found on arrival that I could not drive onto the field. Luckily it all worked okay but I did feel a little stressed at the time. If I cannot lead the animals but have to herd them, they could escape or injure themselves.

New owners usually need support once they have their animals. They may find it difficult to put on head-collars, they don't know how to get the animals into the catch pen. They want to be shown how to trim the toenails. I usually expect to make to two or three visits. I remember to take rope, head-collars, lead-ropes and toenails trimmers. Customer may have purchased all of these, but may not have them to hand when I turn up. Taking my own can save a lot of time.

The rope is useful for making a kind of funnel into the catch pen. Tie one end of the rope to the pen, as the alpacas move into the funnel bring the rope around behind the alpacas encouraging them into the pen.

I take care when using a rope. If a group of alpacas run and take the rope with them, I just let go. I always keep the rope away from my feet, legs and body, holding the rope end with just one finger and thumb.

20th Oct

My stud male Chester lives next to the females. He is our first male and does not like to live with other males. He is very polite with people and will very gently take a piece of carrot from the hand.

I have noticed that he has fallen in love with our dog Sasha. He watches her playing in the garden which is next to his paddock. He stands by the fence waiting for her to come over and give him a kiss. She puts her head up and wags her tail. He puts his head down and they rub noses.

Sometimes he even prances about a bit to entice the dog to 'play'. So it is officially love, at least one kiss a day. Sasha likes Chester, she would like him to play, however she has given up with taking him a stick as he never wants to throw it for her!

Our pet dog and relaxed cria

When the babies are being weaned I do leave Sasha out near them so they get used to dogs and this works very well.

25ᵗʰ Oct

I am going to move Chester to a paddock away from a couple apple trees. I have a lot of apples on these trees and I think all the apples may fall at once if we have a bit of wind. I do not want Chester to suffer from apple overload.

I will also do some routine chores. I tend to put a little hay out each day for the alpacas and sometimes they don't eat it all. So today, I will rake-up all the spoiled hay.

Eye Discharge

Chester has a runny discharge from coming from one of his eyes. I try washing with eye with salt and warm water. I check later that day and there is more discharge. Washing the eye can flush out minor debris. Again I washed his eye and he continued to have a little problem. He seemed very unconcerned about his eye and did not rub it at all.

On the advice of other breeders, I buy eye ointment from the pharmacy. This contains a very mild antibiotic. I ask my husband to hold Chester's head and again washed his eye with warm water and salt solution. He is worried about what I am going to do but really likes the sensation of the warm water. Then I applied the ointment. Fortunately no dramas and the following day the eye has cleared up.

I have successfully treated the minor eye ailment, if the problem had been more severe I would have called the vet.

26th Oct

It is time to train the alpacas to use the field-shelter. In preparation, I have dragged my wooden hay feeder onto the hard standing in front of the stables. I had tried just putting a slice of hay out on the ground for the alpacas to eat. But a lot of hay was being wasted and one alpaca claimed all for itself and drove the others away. The hay feeder works well as it is square so alpacas can feed on all four sides. Wastage is also reduced as the alpacas cannot roll in the hay or walk on it. The next stage is drag the rack into the open fronted stable, quite near the door but where the hay will be out of the rain.

27th Oct

This morning I went down to the paddocks and checked the hay rack. Guess what, all the hay had gone, so the placement of the rack under cover is a success. There was a tiny bit left which I stuffed into a tub for Chester. I leave the tub on its side, to keep the worst of rain off, in his catch pen. I check the next day and Chester has eaten all the hay and there was very little wasted so this was a success as well. My alpacas need to be taught by small steps. Last winter I just put hay in the open stable for them and they completely ignored it.

Feeding Groups

Very large alpaca farms may keep animals of the same age, sex and reproductive status together. These groups get on best with their contemporaries for example non-breeding young males like to play together.

The members of each group should have the roughly the same dietary requirement. For example gelded males require only hay, grass and a tiny bit of grain in the winter. Pregnant females with babies at foot require more calories, coarse mix, a lot of hay and soaked sugar beet in winter. Thin females and their babies may also be penned together indoors on cold nights.

Trimming Toenails

In the summer toenails become incredibly hard and that makes them very difficult to trim. I wait for rain, the water works wonders in softening toenails.

A lady, who had her own alpacas, came to visit and learn how to trim toenails. She had tried to trim nails on her own and cut straight across the nail through the quick or nerve. She said there was a lot of blood. I told her the trick to cutting nails is to trim the very tip of the toenail and then carefully trim the sides of the nail cutting around the quick.

I often let the animals eat a little hard feed as I trim. This helps the animals to relax and makes it a more pleasant experience. I get a helper to hold the head of the alpaca. Then I put my hand on the shoulder of the alpaca and slowly run my hand down the leg to the foot, then they usually allow me to lift the foot with no dramatics.

Just quickly grabbing the foot is a bit scary for the alpaca. I think because they are a predated animal something unexpectedly getting hold of the foot is usually bad news. The other thing is to keep the foot of the alpaca quite low so the ankle joint is not flexed too far.

If the animal is still trying to jump around as I trim the nails, I stand them against a solid fence or wall that they can lean against. I use my body to push them gently against the object so they are secure and it is easy for them to balance on three legs.

30th Oct

Collect the dung and find worms in it. I think Penny may have worms. I treated her on the seventh of this month so I need to check with the vet if this is okay or whether I need to worm her again. The vet did say I could worm every six weeks.

Penny and cria

November

Bonfire night the sky is alight with distant flashes and we can hear bangs. I check the alpacas and they seem very chilled out. My alpacas are used to loud noises. We do have a lot of bird scare cannons on local farmland. Some are right next to our neighbour's fence. The alpacas take no notice when these go bang.

Shooting pheasants and rabbits is also a popular local sport. Another noisy thing is, we live near the river and the military use this landmark for helicopter training. They come along nearly every Tuesday and practice low level flying, in the daytime and at night. The alpacas take no notice of any of these things and carry on with their day in their chilled out way.

I feed the alpacas and find a dead tree has fallen in the high winds last night. Luckily it missed both Chester and the fence when it came down.

I will add 'check and remove dead trees' to my list of summer jobs. I was lucky the tree didn't take the fence with it as we know that alpacas are curious and like to go places!

It is time to trim the alpaca's toenails. I find it is much better to do this job frequently. Leave the nails to grow long and they 'close-up' over the quick or nerve. The nail bends over as it continues to grow. This makes trimming very difficult. Trim often, just pick up the foot, give a quick snip and put the foot down again, easy!

Curious alpacas like to go places

Toenails do wear down naturally on the Altiplano, where it is rocky. No rocks here in East Anglia and toenails grow long. One thing we can do is put down concrete pavement slabs, where they feed or stand gazing over the gate. This helps to wear down the toenails and minimise poached ground.

15ᵗʰ Nov

I have been picking up fallen apples. I put the very best apples, with no blemishes, in boxes to store for the winter. Last night we had a very hard frost so the apples which are left on the ground I will feed each day to the alpacas while they remain good.

Feeding my home grown apples has made a difference to the condition of the alpacas. Two slightly underweight mums have put on weight. The only change to their diet is two apples a day!

November

Our alpacas love their apples. I scatter them on the paddock in the morning and the alpacas choose their apple and have a little bite at them. The hardest bite is the first one because the apple is round, smooth and rolls out of the way. From then on eating them is plain sailing.

24th Nov

The weather forecast says temperatures will go down to freezing this week with a chance of snow. I am now feeding my male alpaca two buckets of hay a day plus apples. He looks enormous, summertime 'the living was easy'.

I have two pregnant females with four and six month old babies at foot. I am feeding them one scoop of alfalfa, some soaked sugar beet and half a scoop of alpaca mix with apples and hay twice a day. I don't feed too much grain because one of the ladies seems to get tummy ache if she eats a lot of grain. She lays down after about only one minute of feeding but does this every time if she is eating a diet which is high in grain.

27th Nov

Everywhere is sparkly and silver. Very heavy frost and one degree is forecast for the day. I moved the females back to the stable paddock, which is very sheltered with buildings and trees on two sides.

I have to use boiling water to defrost the soaked sugar beet that is in the feed store. After I have mixed the feed it feels tepid. I have tried giving them a warm meal but they will not eat it. I leave a whole bucket of alfalfa to eat through the day, in case they are hungry.

I put them down a bed of hay in the stable. Good hay should not be dusty, this can mean mould spores. It should smell sweet and obviously have no dark patches. I have started to give Chester a little hard feed and hay now it is definitely down to freezing.

The forecast is for a cold snap lasting two weeks. I still have ten bales of hay in stock and loads of alfalfa and sugar beet. Our next load of hay is booked in to be delivered early next month. I have run out of alpaca coarse mix. I get ten bags of feed from our local mill and they give me a twenty five percent discount. But the delivery price would eat up the discount so I am waiting for my husband to have some free time to go up to the mill with our trailer.

The mill is very efficient and they have all the bags loaded onto a pallet. The fork lift driver then just puts the pallet straight onto our trailer. It just goes to show it is really good to have plenty of food in by early November as the cold snap can come sooner than expected.

Winter Feeding

When it is very cold I feed the alpacas as much hay as they can eat. For the females I also leave out extra alfalfa in case they feel like a snack.

Adult male alpacas can get fat, if living on good grass all summer without the strains the females have of growing and feeding babies. The top studs work hard, with lots of mating, but males on the smaller farms have less work. And of course geldings have nothing to do but eat.

I have been asked for the exact temperature when I start giving the boys some alfalfa or coarse mix. When the temperature gets down to freezing I normally start feeding the males. I also check the quality of the grass they are eating. The nutrition in the grass reduces with falling temperature and reduced daylight hours. Grass with yellowed ends is definitely less nutritious.

I feed a bucket of hay a day in early November. As the days get colder I increase this to two buckets a day per animal. If they don't need the hay they will not eat it. When we get to nights going down to sub-zero temperatures then I do start feeding alpaca coarse mix with alfalfa. As always check the back of the animal every two weeks. If they feel like a table top they are too fat. If the spine feels sharp they are too thin.

28th Nov

The alpacas are wearing a frosty overcoat this morning. They have spent the night outside. I check the backs which feel very nice and rounded. They have steadily put on weight over the autumn and are now at optimum weight ready for the winter.

Frost on her back

Alpacas can loose and gain weight quickly. Breeding females need close attention to keep their weight correct. Dramatic weight loss can occur during the first two months of feeding a new cria.

20ᵗʰ Nov

It has snowed and the females are all dry this morning and no snowy overcoats. They are not very hungry for breakfast. I think they had been busy eating hay all night indoors.

Chester goes back to his paddock that has a field shelter. I put him on the halter to move him. Now sometimes Chester is tricky to catch, he swings his rear round when I try to get hold of him. When

I do get a hand on, he jumps sideways to try and get away. Today he was a perfect angel and I can only put it down to more love. I have been spending time near him watching the world go by as he eats his breakfast. I have been taking him an apple each day and calling him over to eat it. This little bit of extra effort has paid off as he let me catch and put his head collar on easily and he walked like a star on the lead rope.

Snow and alpacas have spent the night inside

Jackdaw

We have a Jackdaw living in our garden. He has one wing folded at an odd angle and seems to just hop around on the ground most of the time. He sits and waits for me to go and feed the alpacas. Then he hops into the end of the feed trough and eats at the same time as the alpacas. They don't seem to want to chase him away, even from the yummy grain!

30th Nov

A client telephones, he has a sick alpaca. He is feeding a dried rye grass product instead of hay. His vet has diagnosed rye grass staggers. The cure is a change of diet and stable rest. His alpaca makes a good recovery.

Jobs

The next thing to do is toenail trimming and halter training. Nina is coming up to six months which is the optimum age to start halter training. The stable paddock has the best size pen and the females are already on this paddock at the moment.

Halter Training

I start by feeding in the catch pen regularly. When the youngster is relaxed with this I can start training. I have an even smaller pen added on to the feeding pen. I gently move the alpaca into the smaller space. Then gently catch them and hold by the neck.

Next hold the halter below the alpacas chin. One hand holding each side of the halter, with the alpaca still encircled by my arms. Slide the halter quickly up over the nose.

Young alpaca having a halter put on for the first time

And quickly buckle-up the halter. The alpaca in the pictures has never worn a halter before. Because we have prepared him by feeding in the catch pen and we have previously handled his head, he is happy to be near us and have the halter put on.

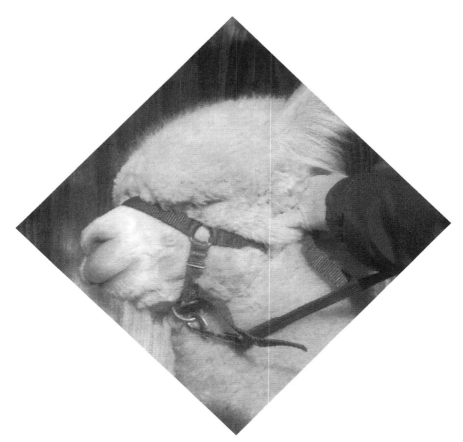

Fasten the halter behind the ear

Move gently and slowly, the young alpaca is watching me and reacting to my body language. When in the small pen I will follow the alpaca around the pen.

Alpaca watching my body language

And then I gently ask for a step or two, by putting a gentle pull on the lead rope and then releasing the pressure when the alpaca takes a step.

Gently ask alpaca to take a step

I normally only spend five minutes on each training session. Young alpacas have a very short attention span. For the next lesson we will repeat todays work and also move around the larger pen.

Then we go on a walk following the Mummy alpaca who is being led by an assistant. We repeat these walks a couple of times and then our alpaca is just about halter trained.

Alpacas do like to be led in pairs. If they are unsure about moving forward it is best to take a step backwards and get them to move forward away from you.

This young alpaca is now happy to wear a halter.

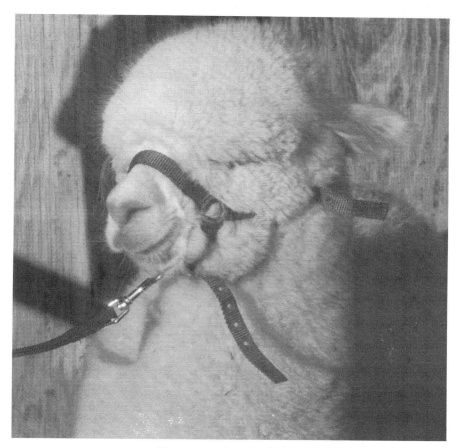

Happy wearing the halter

December

Very cold again today, we have had a little more snow overnight and the wind chill is making it feel like the arctic. The females have been in their shelter eating hay. They don't seem to be particularly cold or hungry. Chester has been standing where he can see the females rather than in his shelter and he is shivering so I give him a bit more breakfast than normal.

6ᵗʰ Dec

Another very cold night, the thermometer says minus three this morning. The females were not frosty on their backs so they have had the night in their shelter. No 'green patch' on the ground in Chester's paddock, so he was in his shelter last night too.

Penny is still feeding her cria which is just over six months old and looks almost as big as her! I do need to wean the cria, the problem is I need a pair or more to put together as company when weaning. The other cria is too young at four months old. So I will have to leave this till February.

30ᵗʰ Dec

It has been very cold running up to Christmas. The thermometer has read minus six at breakfast time. I take out a watering can of hot water to defrost the water troughs.

Breaking the ice works okay if the temperature rises above freezing in the daytime. Otherwise ice refreezes thicker than before. The alpacas seem to be coping very well with the cold weather.

Making Money from Fleece

How to make money from alpaca fleece?

Sell the fleece, just as it was sheared from the alpaca, to hand spinners. I sell it locally by word of mouth and also sell on the internet.

Learn how to spin the fleece into wool. Once you have learnt, it is easy to do and very relaxing. Sell the home spun yarn to hand knitters who find it very desirable.

Learn how to knit and make knitted garments. Hats and scarves sell very well at local fairs.

Needle felting is another craft that is easy to learn and lots of fun. Then sell your felted art creations. Bags, slippers and lovely table mats can be made using the 'wet' felting method.

Take your fibre to the mill and get it spun into knitting wool. If you have 25kg or more of the same colour fibre you can send it to one of the mini mills for processing.

For single fleeces or amounts less than 25kg you will can send your fibre to one of the mini mills specialising in small amounts. Processing costs for fleece vary upwards from £30/kg of received weight.

The greatest return from your fibre is to manufacture it into quality garments, but this does requires a large financial investment and do remembers that fashions change very quickly!

How to Sell Alpacas

I start by letting people know I have alpacas for sale. Some alpaca breeders have an open day or take alpacas to local summer shows and fairs. I use my website to advertise my alpacas. There are general alpaca websites and for an annual fee they will advertise your farm and alpacas for sale.

I create a data sheet for each animal that my customer can take home and study. I include statistics on the fleece, pedigree and show results. I write a brief description of each animal's personality and qualities, for example: good mum, easy to handle and halter trained. Finally, I include a gorgeous picture of the alpaca and the selling price.

When people come to your farm to look at alpacas they may want to buy, so have the printed descriptions ready on your animals for sale. Your customers may not be able to decide on an animal immediately. They will find it very helpful to take away information on the ones they are interested in buying. You will find they may even telephone later having made the decision!

In the spring, maybe have a day when you access your stock. Prepare a form with a column for each animal. Your form could look like this one, or you may wish to add some of your own categories.

I give each animal a number for the fineness of their fleece, the density and crimp. I decide on the numbers by feeling and looking at the fleeces. I then look at the conformation of the animal and again assign a number.

December

Using this information I decide which animals to sell, who I want to keep, assign a price and write out the description for my customers.

I did wonder what I would do with boy babies. But once they are weaned they sell very easily as pets. I got together with other small breeders in order to make sure each boy alpaca would have two alpaca companions in his new home. Do remember to tell their new owners that males should be gelded between fifteen to eighteen months of age, ideally in the winter months, when flies will not get onto their wounds. If left entire they will start to fight with each other and be more challenging to handle.

	Alpaca Name	**Lucky**	**Purdy**	**Mimy**
Fleece	*Fineness*	9	7	7
	Density	7	5	4
	Crimp	7	6	7
	Colour	White	Black	Brown
	Solid Colour	Y	-	-
Conformation		8	7	9
Proven Breeder		Y	-	Y
Good Mum		Y	-	Y
Easy to Handle		-	Y	Y
Friendly		-	-	Y
Halter Trained		Y	Y	Y
Notes		-	White Face	-

Stock assessment form

Fun with Fleece

My mum is brilliant at knitting, speedy and accomplished she quickly made lots of jumpers for my brothers and I. Going with her to the wool shop, was a real childhood treat, all those balls of wool in beautiful colours. Mum taught me to knit too and to understand how to read a knitting pattern.

When my alpacas had their first shearing, by lucky chance, my daughter's music teacher showed me her jumper. She had spun the wool and knitted it. I arranged a spinning lesson and I borrowed her spare spinning wheel. It did take a lot of practice. Then I purchased a spinning wheel with two treadles. Now I pedal gently with two feet, I find spinning wool a very pleasant thing to do!

My tip is to spin all the wool for a jumper and then knit the jumper. I used to spin a bit then knit a bit and found the gauge of the wool had changed by the time I reached the last sleeve!

Now I always knit a test square so I can calculate how many stitches and how many rows are needed for the jumper.

I also take some fleeces to my local mini-mill. I have them washed and carded, and turned into a 'bump', a long thin carded strip about 1cm wide wound in the shape of candy floss! My friends say, spinning from a bump is like spinning silk. I also have my fleece processed into hand knitting wool and machine knitting wool. I take the saddle or back area of the fleece to the mill. I remove any coarse areas with guard hair and any foreign bodies that have been caught in the fleece.

White fleece can be turned into all the colours of the rainbow, by using dye. I have used proprietary wool dyes, fabric and cold dyes, with good effect. After using dye, blend out any dark or light bits when carding the wool before spinning. Spin different colours onto two spindles and then spin these together for two-ply, two-tone wool.

Cria inspecting the knitting!

Tie the fleece in an old 'plain mesh' piece of net curtain, to avoid it turning into felt. Lift the fleece in and out of the dye using only the 'knot' of the net curtain. Rinse the fleece by placing in a container, and leave under gently running water. Carefully squeeze out excess water and put the fleece out to dry in the sun.

Knitting wool can be dyed. Wool wound in a ball can be rewound as a skein. Tie thread around the skein in four places and then begin dyeing, gently stir in the dye to get an even colour.

Felting is an easy way to make scarves, bags and slippers, it is a little like magic. Start with a big pile of clean carded fleece. Lay it out in three layers, alternate the fibre direction on each layer. Apply hot water with clothes washing powder and cover with net or 'bubble wrap', then rub with pressure. Keep rubbing until the fibres knit together, use more hot water and washing powder if necessary. Then rinse in very hot and very cold water, this is called shocking, used to strengthen the felt. Repeat these processes until the felt shrinks to the correct size.

Alpaca *felt slippers*

I have made slippers and bags using a plastic former. To make the former for a slipper, draw around your foot on a piece of paper and then enlarge this by two inches all round. Use this as a template to cut a former out of stiff plastic. If you don't have any plastic use cardboard and cover in cling-film to stop it getting soggy. Or you can use an old plastic beach shoe as a former. Just tie it in a plastic bag.

Lay out fibre for one side of the slipper apply some hot water mixed with washing powder. Lay on the former. Fold in the edges over the former. Now put on fibre for the top side of the slipper and again apply washing powder and hot water. Fold under edges to the bottom side of slipper. Cover with bubble wrap and rub with pressure. When the fibre has knitted together cut a vertical slit, this is where the foot goes into the slipper, and remove the former. Turn slipper inside out and keep rubbing and rinsing until it reaches the correct size.

Alpaca Time

I have learned so much and I am enjoying passing my knowledge on to others. Alpacas have changed how I spend my time. Breeding and routine care of the alpaca is a very satisfying occupation. Alpaca fibre is amazing, soft like cashmere, strong like silk and warmer than sheep's wool. I really enjoy all things that I do with the fibre, including selling at local craft fairs.

I like talking to people about alpacas. As well as the scarves, knitting kits and felted items, I always take some raw fleece with me. People love to plunge their hand in and feel just how soft it is!

Enjoying alpacas

Alpaca are charming animals to spend time with. They are gentle and curious. They can have a calming effect on people and they require very little work to look after.

The babies are gorgeous to look at and playful. In the morning and evening young alpacas perform a 'cartoon' like running, called spronking, which is skipping along with a jack in the box type jumping motion. All four feet touch down then all four feet off the ground. Sometimes the adults join them in a mad dash around at sunset.

Gorgeous cria

Alphabetical Index

Printed in Great Britain
by Amazon.co.uk, Ltd.,
Marston Gate.